"That's it! It will be 'OPERATION BUMBLEBEE'." Merle Tuve had just spotted this aphorism hanging on the office wall of Capt. Carroll L. Tyler:

THE BUMBLEBEE CANNOT FLY

According to recognized aerotechnical tests, the bumblebee cannot fly because of the shape and weight of his body in relation to the total wing area. BUT, the bumblebee doesn't know this, so he goes ahead and flies anyway.

Well aware that this new undertaking to develop a supersonic guided missile for the Navy would face impossible challenges, Dr. Merle Tuve named the project "Bumblebee".

RAMJET ENGINE

"The Ramjet is the basis for all our jet aircraft. If you consider our dependency on jet transportation, Topsail Island's contribution to today's technology and ability to travel globally is significant."

[Lt. Commander Tad Stanwick]

The Ramjet engine was proved here on Topsail Island. This development is as significant to jet travel as the first-flight, venerated at Kitty Hawk, is to propeller flight.

GUIDED MISSILES

"Another triumph was the successful missile-borne radar beam riding tests. This was an early proving of feasibility of control and guiding of missile flight." *[Lt. Commander Tad Stanwick]*

"This project gave the Navy enough knowledge and know-how to put the Terrier, Tartar, and Talos missile systems aboard U.S. Warships with the capability of destroying enemy airplanes at a range far beyond that of naval guns." *[Lt. Bruce L. Goodwin]*

CONTENTS

PICTURES - IN ORDER PRESENTED   CREDITS

Front Cover Composite	JHU/APL*
Aerial View of Firing Point	JHU/APL
View of Firing Point From Range 1	JHU/APL
Assembly Building Today	David Stallman
Control Tower & Assembly Building -	5/57–J.B. Brame
Ramjet Rockets	Historical Society
Weighing Rocket–Back of Assembly Bldg.	JHU/APL
Patrol Boat Inland Waterway	Historical Society
Off-Shore Range Patrol Boat	Historical Society
Launching Pad and Launchers	JHU/APL
Large Rocket Launcher	JHU/APL
Rocket at Assembly Building	JHU/APL
Rocket Into Launcher	JHU/APL
Rocket Final Checkout	JHU/APL
Rocket Blastoff	JHU/APL
Tower Cine-Theodolite Equipment	JHU/APL
Bombproof Room	JHU/APL
Pontoon Bridge - Sears Landing	JHU/APL
Warehouse Building - Surf City	JHU/APL
Back Cover–Map	JHU/APL

*JHU/APL - Johns Hopkins University Applied Physics Laboratory

OPERATION BUMBLEBEE

PREFACE

The concrete structures in our midst have stood the test of time and ravages of storms. They have fueled the imagination of Topsail people and, over time, some of the secrets were revealed, but they were still shrouded in mystery. Discovery of declassified military documents and pictures in the Johns Hopkins University Applied Physics Laboratory (JHU/APL) Archives finally gave us military proof of Topsail Island's missile heritage.

The Historical Society of Topsail Island was determined to preserve these historical structures. In 1993, the Assembly Building, Control Tower, and Tower Number 2 [Queen's Grant], as a group, were designated historical sites in the National Register of Historic Places. Tower Number 2 was included as it most accurately represents a tower in its original state.

The Topsail Beach Economic Development Council officially signed the documents of ownership, in 1995, for the Assembly Building. It is their aim to restore the building as a historical site and provide Topsail Island with a Community Center that can serve many purposes.

This historical recall includes personal accounts, official Navy and JHU/APL descriptions of the project, and actual pictures. Every attempt was made to ensure accuracy of the information. It is hoped that the reader will find this as exciting as I did in the discovery.

The personal accounts by Tad Stanwick and Bruce Goodwin were invaluable and I thank them. Philip K. Albert, Sr. of JHU/APL Archives was instrumental in the breakthrough discovery of pictures and documents, and I owe him sincere thanks. Other Navy and National Archivists also provided information and direction to sources.

David A. Stallman

BIRTH TO DEATH

On the heels of WWII, Topsail Island (known as the sand spit) was seized by the U.S. Government for use as a secret missile test site. Johns Hopkins Applied Physics Lab (JHU/APL) was committed to the long-range role of defense support for the Navy. Their aim was to develop a supersonic missile that could swiftly reach out ten or twenty miles to hit and destroy an air threat. They needed more range than possible at former sites in New Jersey and Delaware.

The project was one of the earliest developments of the nation's missile program. It was managed by the Navy Bureau of Ordnance and guided by the Johns Hopkins University Applied Physics Laboratory. Topsail Island was ideal for such a secret project because of the few residents and limited access via a pontoon bridge.

An existing Camp Davis at Holly Ridge was reworked as a base for 500 personnel. Water and electricity were "piped in" from Camp Davis. Eight observation towers were built along what they called this "sand spit" to house timing instruments and cameras to record performance data. All stops were pulled out and the range, incredibly, was built in one year. This missile test range was to be a permanent installation and claimed to be one of the best equipped fields in the country.

The Assembly Building was a hardened structure constructed for assembly of the rockets. It was carefully designed with copper grounding and lightning rods at corners of the building to prevent explosion. Some 200 experimental rockets three to thirteen inches in diameter and ranging from three to thirteen feet in length were fired in the 1946-1948 period and were dubbed "Flying Stovepipe". It was here through these tests that the ramjet engine proved to be a success. Modern jet aircraft have built upon this original design.

In 1948 it was decided that more range was needed and the weather was not stable enough for precise instrumentation required. The program was transferred to test sites at Inyokern, White Sands, and Cape Canaveral. The land and buildings were turned over to the landowners and salvageable equipment to local town and county governments.

MISSILE SITE

The missile site was principally made up of an Assembly Building, Control Tower, Launching Platform, Bombproof Room, and eight photographic towers. They are all basically intact except for several of the photographic towers.

There were also other support type buildings such as Mess Hall, Sleeping Quarters, and Repair Shop, which were located at the present Breezeway Motel site.

ASSEMBLY BUILDING: This building was used for missile storage and assembly. It is a hardened structure with reinforced walls, built on four feet of concrete secured by 20 foot pilings. Ground-straps on doors were to minimize risk of spark and explosion. It also had a lightning rod at each corner of the building for protection, of which three remain.

FIRING POINT CONTROL TOWER: The control tower sits on a line between the Assembly Building and the Launching Platform. Originally the tower had a control and observation deck on top from which the launchings were controlled. There was two-way communication between the control tower and the photographic towers.

LAUNCHING PLATFORM: The platform was a 75 x 100 x 1 foot concrete slab and now serves as part of the Jolly Roger Motel patio.

BOMBPROOF ROOM: The bombproof room was built of 14 inch reinforced concrete walls with a 4 inch by 3 foot window slit for rocket scientists and technicians to observe rocket firings close-up. Access to the room was through a heavy steel sliding door at the street side. It now is in use as part of the Jolly Roger Motel basement.

PHOTOGRAPHIC TOWERS: The towers were precisely located with distances scientifically derived. They were rigidly constructed of reinforced concrete and built on pilings driven to a minimum of 20 feet and 15 tons bearing. To preclude distortion of tower frameworks due to temperature changes, the sides of the towers were protected with plywood shields. Triangulation of photo equipment from the towers would record the flights over 10-20 miles at speeds of 1500 mph, so accuracy of the data was paramount.

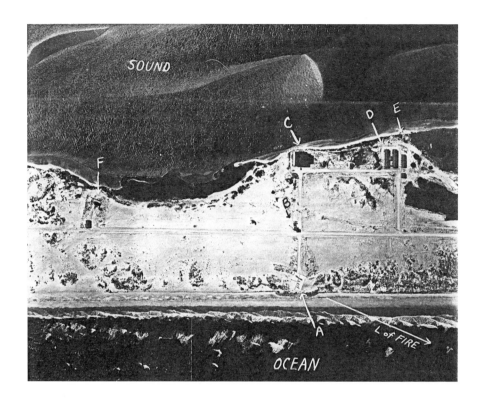

Figure 4.

Aerial Photo of Firing Point. A–Firing Ramp, B–Control Tower,
C–Assembly Building, D–Stock Room, Electrical Shop,
Photo Shop, Camera Shop, and Machine Shop,
E–Range #1, F–Instrument Tower #1.

AERIAL VIEW OF FIRING POINT

Figure 5.

View of Firing Point from Range One. A–Firing Ramp, B–Firing Ramp Bombproof Shelter, C–Control Tower, D–Fueling Shack, E–Assembly Building, F–Stock Room.

VIEW OF FIRING POINT FROM RANGE 1

9

ASSEMBLY BUILDING TODAY

CONTROL TOWER & ASSEMBLY BUILDING Circa 1957

MR. IRVINE B. IRVING - *personal account*

In a *Fayetteville Times* interview, 1978, Mr. Irvine B. Irving, staff member of Johns Hopkins Laboratory, recalled: "We first went down to Topsail in the summer of 1945 to look around and survey the site. At the time, APL was a prime contractor with the Navy in the development of a ramjet propulsion system for supersonic missiles. We moved on the island in March 1946, and lived in the old Army barracks while the firing and tracking systems were being constructed. As I remember, the Kellex Corporation and several other contractors were involved and Navy and civilian personnel numbered as high as 200 at times. The first test missiles were launched around October or November 1946, and from then until the project was moved west in early 1948, we fired some 200 experimental rockets.

"Missiles ranged in size from three to 13 inches in diameter and from three to 13 feet in length and were fired along a range that ran some 20 miles in a northeastward direction."

According to Irving, the testing program produced vital information on guidance systems, aerodynamics, solid propellants, booster configurations, and similar scientific data, all extremely important to the still fledgling missile project.

"Everything went smoothly," he said. "There were a few storms, but no serious accidents, and our work resulted in development of the Terrier and the Talos which were the first supersonic missiles utilized by the Navy."

Nicknamed the "flying stovepipe", the six-inch diameter ramjet was fashioned from the tailpipe of a Navy Thunderbolt airplane. It was powered by a mixture of propylene oxide and the oxygen of the atmosphere scooped into the front end of the open pipe, compressed in the chamber by the speed of the vehicle, and ignited.

Much of this experimental development occurred on Topsail in 1946-47 as the Navy refined its "flying stovepipe", and kept well ahead of the Army, which did not successfully test fire a similar missile (Corporal E) until May 1947.

RAMJET ROCKETS

WEIGHING ROCKET-BACK OF ASSEMBLY BUILDING

14

TAD STANWICK - LT. COMMANDER

- *personal account*

Mr. Stanwick owns a house on Topsail Island. Commander Stanwick was the Navy's top technical and scientific liaison officer on Research and Development of the Terrier and Talos Guided Missile Programs. He was at the forefront of "Operation Bumblebee" and remained the Naval officer in charge for the duration of the project. He shares his perspective and recollections.

The Bureau of Ordnance had the facility built, and was remarkably able to put such priority and resources on the project to complete the facilities in about one year. They brought in water and electricity from Camp Davis. Stanwick had towers and roads built and obtained fire trucks, etc. The towers were precisely located at documented longitude and latitudes with distance between towers scientifically derived. Photo theodolite cameras which would do telemetering of rocket firings were installed. Data accuracy was paramount to the success of the project.

The pontoon bridge had to be rebuilt. The original one was an atrocity on which you could load one or two trucks, turn winches on until it arrived across the channel, and drive off. The Corp. of Engineers built a more permanent bridge.

Some 580 APL, Navy, and Marine people were assigned to the "Bumblebee" project. Navy radar technicians and scientists were all under Stanwick's security responsibility. They had two patrol boats operating in the sound and a 65 footer in the ocean based at New River Inlet. Their prime responsibility was to monitor fishing boats. They would let fishing boats work but would warn them of firings. Of course suspicious activity considered a threat to project security would be reckoned with too.

It was pretty desolate at that time, particularly at the north end. Commander Stanwick said, "I would often take an inspection trip in a jeep from the north end to the south end of the island. It was deserted at the north end–

scary–not a soul there. With the secret equipment used on this project nobody was allowed on the island without special security."

The Research and Development was done in Silver Spring, MD. APL would test components. Solid rockets as boosters came from Cumberland, MD. All parts were shipped here, assembled, and fired. They were two stage rockets. Solid rockets would boost it off the launch platform and the ramjet would take off away from it at supersonic speed. It is hard to estimate the number of firings. It would take about two to three days to get one rocket ready for firing, plus dummies were fired to test equipment.

The Sand Spit [Topsail Island] site was fully intended to be a permanent installation. The government had obtained 99 year leases where needed and seized the land. The project moved from Fort Myers, Delaware, because of the need for more space. Their mission was to get the ramjet up to supersonic speeds and demonstrate that it developed thrust. Booster technology was developed here and the island's long shoreline provided 26 miles for the range. Up to then, the ramjet would light and burn but they couldn't prove the theory. With their instrumentation they proved the ramjet could develop and maintain thrust.

The ramjet is the basis for all our supersonic jet aircraft. If you consider our dependency on jet transportation, Topsail Island's contribution to today's technology and ability to travel globally is significant.

Another triumph was the successful missile-borne radar beam riding tests. This was an early proving of feasibility of control and guiding of missile flight.

A lot of resources had gone into making this a permanent installation. It was discovered that the weather was not conducive to good instrumentation and the range length was proving inadequate. There were other areas available that would prove to be better suited. Stewart, Florida (Cape Canaveral/ Kennedy), Inyokern, California, and White Sands, New Mexico, all got pieces of the project when it was dismantled.

The site facilities were left intact and turned over to the Navy Yards and Docks to dispose of by selling or giving to bona fide town, county, or state government agencies. The only things sold were piping and supplies. All else, such as electricity and water systems, went to the respective towns/ counties. Mayor Elvie White of Wilmington begged them not to remove the facilities. Many people wanted to dig up pipe for salvage. A lot of stuff went to the City of Wilmington, fire trucks went to Holly Ridge and Hampstead, and bulldozers and road scrapers went to State roads people.

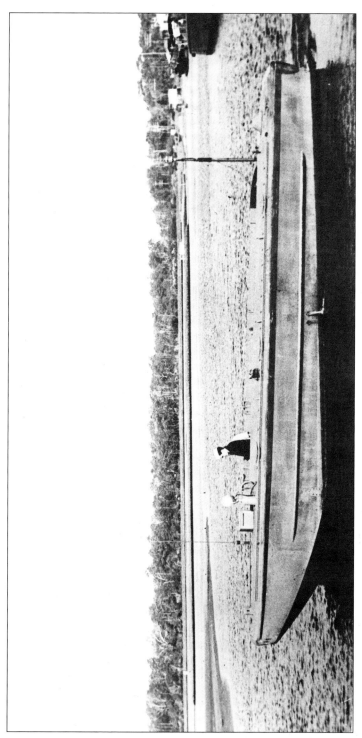

Figure 14.

Inland Waterway Patrol Boat (LCPL).

PATROL BOAT INLAND WATERWAY

OFF-SHORE RANGE PATROL BOAT

18

MR. BRUCE GOODWIN -
LT. KELLEX GUARD- *personal account*

Mr. Bruce Goodwin, a retired Lt. of the Kellex Guard force shared his recollections. "The first time I visited Camp Davis was in April 1946. The firing point, concrete towers, and assembly building were under construction. A year later, I was employed by Kellex as Lieutenant in their guard force. The missile range was already in operation."

Kellex was the prime contractor and was responsible for construction work and for overall security. They even had two patrol boats operating in the sound and a 65 footer in the ocean to monitor fishing boats and keep the project secure from access by water.

Crates of missile parts were shipped from the JHU/APL location at Silver Springs, MD, by railroad under guard. Kellex Corp. would dispatch technical and guard personnel to transport the missile parts from Wilmington to Topsail.

The concrete towers had two-way communication with the control tower at the firing point. Their job was to photograph the missile as it roared down-range and check it out by telemetry. The firing point control tower was located between the assembly building and the launching platform. It had a control and observation deck on top from which the launchings were controlled. The bombproof room had 14 inch reinforced concrete walls with a 4 inch by 3 foot window slit for rocket scientists and technicians to observe rocket firings close-up. Access to the room was through a heavy steel sliding door at the street side.

On a firing day, the solid propellant powder was loaded into the missile. It was manually trundled to the launching platform on a rubber-tired dolly to minimize risk of explosion. The launches were crude ramps made of wood and a steel half-casing about 15 feet long set up at an angle. Once the missile was in the launcher, it would go through a series of checkouts.

If the checkout was not successful, the missile would be carefully returned to the assembly building for the night and extra security precautions were in effect. Sometimes it would take several days to launch one.

The first stage solid fuel rockets would boost the rocket off the launch platform and the ramjet would take off away from it at supersonic speed. These launches were scary affairs. Being this new, it was uncertain what would happen at times. Rockets went awry and did crazy things at times, so safety was paramount.

Once Bruce was in charge of security near the launch control tower. As he took up his post, the range officer bellowed at him through a bullhorn to move out of the area. He was chagrined that he was treated in such an undignified way. After all, this was his post. As he vacated post, a maverick rocket crashed and exploded right where he would have been. The safest place to observe from was the bombproof room. "Accidents were few, though, considering the high explosives used. I guess we were cautious because we were scared of it.

"It should be pointed out that this project gave the Navy enough knowledge and know-how to put the Terrier, Tartar, Talos missile systems aboard U.S. Warships with the capability of destroying enemy airplanes at a range far beyond that of naval guns."

LAUNCH PAD AND LAUNCHERS

LARGE ROCKET LAUNCHER

ROCKET AT ASSEMBLY BUILDING

ROCKET INTO LAUNCHER

ROCKET FINAL CHECKOUT

ROCKET BLASTOFF

TOWER CINE-THEODOLITE EQUIPMENT

BOMBPROOF ROOM

DOCUMENTS - Navy Survey Report

- April 26, 1946

PONTOON BRIDGE

The "Sears Landing" bridge was critical to the functioning of this project. While it served a purpose of security by making access difficult, it was woefully inadequate for Bumblebee. A Navy report "Survey of Camp Davis, NC for use as an East Coast Guided Missile Site", reads: "Existing 90 foot barge, equipped with winch and cable arrangement for opening-closing and approach lift will be abandoned, since this portion of the bridge is in very poor condition.

"A new barge, operating on the present fixed pivot swing principle, is to be constructed of 54 units of the Navy's 5 foot x 7 foot pontoon barge equipment. Swinging power will be provided by a Navy 'Sea Mule' outboard power unit, with another as a standby. Barge approach runs are to be counterbalanced and elevated and lowered manually. Present cables strung across the channel, which must be dropped to clear all waterway traffic, will be eliminated. The repaired bridge will be maintained in closed position."

[This bridge remained in use until 1955 when the swing bridge was opened.]

PONTOON BRIDGE – SEARS LANDING

WAREHOUSE BUILDING – SURF CITY

DOCUMENTS - Navy Department Bureau of Ordnance

In a Navy Department Bureau of Ordnance document dated May 3, 1946, the proposed Camp Davis, NC East Coast Guided Missiles Test Range (Project No. Ord 659) was described.

The Secretary of the Navy approved retention of leases on the Camp Davis, North Carolina, site for this Bureau's use in connection with the development of guided missiles.

Initial experimental instrumentation of prototype work was first flight tested at Island Beach, New Jersey, and is continuing on an interim basis at Fort Miles, Delaware.

It will be necessary to furnish complete range instrumentation, together with modification of existing facilities and installations at Camp Davis. This will include provision for housing and subsisting 250 personnel. It is estimated that the following breakdown of the proposed installation will fulfill existing requirements for operation of the East Coast guided missiles test range. The following descriptions were among the requirements:

– Explosion proof assembly shop. 60' x 80' explosion proof building with reinforced concrete deck supported on piles. Walls of building for 5' of height to be of reinforced concrete. Next 10' of height, walls to be of concrete block. Steel beam truss supported roof made of corrugated transite. Monorail and chain hoist to run 80' length of building. There is to be a 25' section partitioned off at one end as explosion proof room. Building to be air conditioned and heat controlled for 70 degrees constant temperature. Four steel doors, 2 double, 2 single. All electrical fixtures explosion proof. A 10' platform is to be along one 60' side and one 80' side of building, outside, with a 20' wide shed overhead.

– One bombproof shelter 10' x 10' x 10' reinforced concrete 1' thick by launching ramp.

–9 Stable platform foundations for instrument mountings. Pile formation supporting a 30 ft. square platform of reinforced concrete beams with reinforced concrete between the beams. On top of this platform, a 15 ft. square structure, 35 ft. high of reinforced concrete beams with concrete blocks between beams. There are to be a total of 4 decks for placement of instruments; 10 ft. between all decks. Top deck to be open, with 5 ft. high rail of concrete blocks, surrounding.

– 1 Stable platform foundation for mounting launching ramp. A reinforced concrete platform 300' x 50' x (approx.) 1'.

DOCUMENTS - Navy Press Release

NAVAL ORDNANCE TEST FACILITY–
PRESS RELEASE - JAN. 1948

A construction project in which resourceful salvage measures overcame extreme shortages of materials was finished recently at Camp Davis, Holly Ridge, NC, by the Navy's Bureau of Yards and Dock for the Navy's Bureau of Ordnance.

The two-phase contract called for the construction of the East Coast Test Range along the Atlantic Ocean Beach, near Camp Davis, and construction of housing facilities in the camp itself. Salvage work was accomplished at both locations.

Camp Davis was originally constructed by the U.S. Army, starting in 1940, and reached an eventual capacity of approximately 70,000. After the war the Army turned the camp over to the U.S. Marine Corps, which used the camp for training and also for several months as a separation center. The Navy assumed possession 1 June 1946.

Since the present Navy plans call for a complement of four to five hundred persons, only a small portion of the central area of the camp will be used. Outside the central area, some 700 camp buildings of Army standard design "semi-permanent" frame construction have been declared surplus. These have been transferred to the FPHA which has dismantled many of the former barracks and mess halls and shipped them to the Cleveland and Washington areas for re-erection and conversion to GI Housing at college campuses.

Among the structures being continued in use by the Navy are: Administration Building, Central Steam Plant, Telephone Exchange, Family Quarters, Community Center (former Officers' Club), Gymnasium, Bachelor Officers' Quarters, Barracks and Mess for enlisted personnel, Dispensary, Hospital Wards which have been converted into Residential Apartments and Guest House, and various Shops and Warehouses.

The range facilities were constructed on a sand-spit beach along the Atlantic Ocean, approximately 20 miles long and a quarter to half mile in width, and located five miles from the main camp. The range facilities comprise a Launching Platform, located at the south end of the beach with adjacent shops and services, and eight Observation Towers dispersed along the length of the beach.

Communications for the range required the construction of 10 miles of road over the sand dunes and 20 miles of transmission pole line carrying a 7.2 kva power circuit, six synchronization circuits, and 24 communication circuits. A four-inch water line was also constructed to carry service from existing mains at the center of the beach to the facilities at the south end.

Of the range facilities constructed, five were existing frame structures that were dismantled in camp and re-erected at the new site. These were a garage, 26 by 79 feet; a Spare Parts Storage Shed, 26 by 46 feet; a Welding Shop, 26 feet square; and two Instrument Repair Shops, 25 by 110 feet each.

An unusual design feature encountered in this project was the requirement for extreme rigidity and stability on the part of the Observation Towers. These 30 foot high towers serve as platforms for instruments which photograph and locate by angular measurement the positions in space of missiles 10 to 20 miles away, traveling at speeds of 1500 mph. To ensure accuracy in results, it is essential that these instruments be fixed and not subject to movement due to tower vibration or sway. To obtain this stability, the towers were constructed of reinforced concrete and were placed on the centers of 30 by 30 foot concrete slabs supported on creosoted piles driven to a minimum depth of 20 feet and 15 tons bearing. To preclude the possibility of distortion of tower frameworks due to temperature changes, the sides of the towers were protected with plywood shields to prevent undue increases in temperature due to direct sunlight on the frame.

Construction work was performed by George and Lynch, General Contractors, of Wilmington, Delaware. Plans and specifications for the project were prepared by the Kellex Corporation, a firm of civilian contractors associated with the Applied Physics Laboratory of Johns Hopkins University and under whose technical direction Kellex is also operating the range for the Bureau of Ordnance.

DOCUMENTS - Johns Hopkins University Applied Physics Laboratory

CINE-THEODOLITE SYSTEM

JHU/APL documents explained that a theodolite system at Naval Ordnance Test Facility, Holly Ridge, was established primarily to obtain velocity and acceleration measurements for supersonic guided missiles. Basically, each cine-theodolite was "zeroed" on another tower about a mile away with each tower "zeroed" on the next. In that way they could be synchronized and accurately record telemetered data. They tracked the flights using dual 16MM cameras. It was designed for Mach numbers to 5, maximum altitudes of 60,000 feet and ranges up to 80,000 yards. Their measurements were from Tower 1 to 5 miles down the range.

Basic facilities in the technical area consist of eight reinforced concrete, stable platforms or towers and three radar observation platforms located along a 20-mile broken "base line". There are also a concrete launching apron with bombproof shelter, a control tower, and a number of secondary instrument emplacements. All significant points have been surveyed by the Coast and Geodetic Survey.

Technical support facilities include an explosion-protected rocket assembly building, assembly building for inert components, photographic laboratory, optical instrument and camera shop, electronics shop, radar testing building, welding shop, machine shop, and stockroom tool crib.

BUMBLEBEE RESEARCH FEATURED AT NEW TOPSAIL ISLAND MUSEUM

Johns Hopkins APL NEWS/June 1994 announced that the Applied Physics Laboratory Archives would provide Bumblebee artifacts, including mock-ups and films showing Camp Davis activities and rocket test flights.

Later, in 1997, Johns Hopkins APL donated a TALOS missile for an outside display. This impressive and deadly missile was never tested on the island, but the proof of rocket viability and early development of guidance systems that made TALOS possible were accomplished here at Topsail.

DOCUMENTS - Navy Bureau of Ordnance

A November 6, 1947, document lists a Plan for Implementing the Deactivation of Camp Davis. The last missile firing will be November 15, after which no further Bumblebee activity will be planned. It also covered disposition of equipment, water and sewer plants, and security. May 1, 1948, was set as a target date for ending the contract.

ROCKET RESCUE FROM THE SEA

After a storm and some beach erosion, a rocket appeared at the ocean edge on Topsail Island. It was during the Fourth of July weekend, 1994, that a 4 foot tube with broken fins on one end was found encrusted with marine growth. It weighed about 100 pounds and it was feared that there were explosives inside. An explosives team from Camp Lejeune declared it harmless. Comparison to pictures from Johns Hopkins APL confirmed it to be a booster rocket from the Operation Bumblebee Rocket Program, 1946 –1948.

The rocket is displayed in an aquarium to preserve it, at the Missiles and More Museum, Topsail Beach, North Carolina. More rocket displays and video at the museum tell the Operation Bumblebee story.

ABOUT THE AUTHOR

David A. Stallman is an author-historian and publisher of articles and books about Topsail Island history. *Operation Bumblebee* was published in 1992, followed by *ECHOES of Topsail – Stories of the Island's Past* in 1996, with a second edition in 2004.
Website: www.davidstallman.com.

His research of Operation Bumblebee enabled the Historical Society of Topsail Island to gain recognition for the Assembly Building, Control Tower, and Tower #2, all listed in the National Register of Historic Sites. The Missiles and More Museum stands as a reminder of the Operation Bumblebee program at Topsail Island, North Carolina.

Museum website: www.topsailhistoricalsociety.org/missiles/default.aspx